LEXINGTON PUBLIC LIBRARY

THE MILITARY EXPERIENCE.
IN THE AIR: DRONES

An armed MQ-9 Reaper unmanned aerial vehicle taxis down a runway in Afghanistan.

THE MILITARY EXPERIENCE
IN THE AIR: DRONES

DON NARDO

GREENSBORO, NORTH CAROLINA

 To join the discussion about this title, please check out the Morgan Reynolds Readers Club on Facebook, or Like our company page to stay up to date on the latest Morgan Reynolds news!

MQ-1B Predator

The Military Experience.
In the Air: Drones
Copyright © 2014 by Morgan Reynolds Publishing

All rights reserved
This book, or parts therof, may not be reproduced
in any form except by written consent of the publisher.
For more information write:
Morgan Reynolds Publishing, Inc.
620 South Elm Street, Suite 387
Greensboro, NC 27406 USA

Library of Congress Cataloging-in-Publication Data

Nardo, Don, 1947-
 Drones / by Don Nardo.
    pages cm. -- (The military experience. In the air)
 Includes bibliographical references and index.
 ISBN 978-1-59935-384-5 -- ISBN 978-1-59935-385-2 (e-book) 1. Drone
aircraft--Juvenile literature. I. Title.
 UG1242.D7N33 2013
 623.74'69--dc23
                                                          2013011594

Printed in the United States of America
First Edition

Book cover and interior designed by:
Ed Morgan, navyblue design studio
Greensboro, NC

Table of Contents

1. Fireball in the Yemeni Mountains — 9
2. A Weapon Whose Time had Come — 19
3. How Drones and Their Pilots Operate — 29
4. Arguing Over the Ethics of Drones — 37
5. The Future Use of Military Drones — 47

Sources — 58
Glossary — 60
Bibliography — 61
Web sites — 61
Index — 62

CHAPTER ONE
FIREBALL in the Yemeni Mountains

MQ-1 Predator

The morning of September 30, 2011, was warm and sunny in Yemen, a Muslim nation that lies at the tip of the Arabian peninsula. In recent years there had been a rise in political turmoil in Yemen. This attracted groups of terrorists to the country, who set up operations in Yemen's more remote regions.

A few hours after sunrise, a truck moved along a dirt road in Yemen's west-central mountains. Soon it slowed down and stopped beside a patch of desert. Several men got out. They spread blankets on the ground and started eating breakfast. They assumed they were alone. But they were wrong. Invisible eyes, thousands of miles away, watched their every move.

Haraz Mountains in Yemen

Yemen shown in red

A TRAITOR TO HIS OWN COUNTRY

Among the men enjoying their morning meal was a forty-year-old American-born Muslim. His name was Anwar al-Awlaki. For several years he had been a key figure in the Yemeni branch of al-Qaeda. That terrorist group had long opposed the United States. It leaders had planned the mass-murders of thousands of Americans on September 11, 2001.

Most Americans were horrified by the 9/11 attacks. But al-Awlaki later claimed the bloodshed they caused was justified. As a young adult, he had moved to the Middle East. From there, he publicly called on Muslims to launch more attacks on America, betraying his home country.

U.S. officials came to see al-Awlaki as a major threat to the United States. Aided by Yemeni soldiers, American forces tried to capture him on several occasions. But he always managed to escape.

Anwar al-Awlaki

THE MILITARY EXPERIENCE.

FAST fact

Anwar al-Awlaki inspired several troubled Muslim men to commit violent acts. One was a U.S. Army officer, Major Nidal Malik Hasan. He killed thirteen innocent people at Fort Hood, Texas, in 2009.

Nidal Malik Hasan

First responders use a table as a stretcher to transport a wounded U.S. Soldier to an ambulance at Fort Hood, Texas, in November 2009.

IN THE AIR: DRONES

A fire bomb explosion

THE SOUND OF DEATH

This time was different. In early September 2011, U.S. anti-terrorist agents got a solid lead on where al-Awlaki was. For three weeks they tracked him. Finally, using high-flying electronic eyes, they saw him eating breakfast beside his truck.

With al-Awlaki was Samir Khan. An American citizen born in Pakistan, he edited al-Qaeda's English-language online magazine. Another of the men was al-Qaeda's top bomb-maker in Yemen. None of them realized the meal they were eating would be their last.

The first sign they were in trouble came suddenly. One of them caught sight of something moving in the sky. He swiftly warned the others. Dropping their food, the men leapt to their feet and ran for the truck. As they started climbing in, a sharp buzzing sound filled their ears. There was no time to locate the source of this noise. A second or two later, the truck erupted into a blazing fireball. For Anwar al-Awlaki and his companions, that noise had been the sound of death.

THE MILITARY EXPERIENCE.

Spotted by Onboard Cameras

The pilots of the drone that killed al-Awlaki knew exactly where he was. This was because they could actually see him. The drone's onboard cameras spotted the truck. They sent images of the men eating breakfast back to the pilots. When the drone was in the right position, its controllers fired the missile. A few seconds later, the deadly Hellfire reached the target. The truck and all the terrorists were obliterated.

INCREDIBLE TECHNOLOGY

There was a reason the terrorists were desperate to flee. They were well aware of what was closing in on them. It was a bomb. To be more exact, it was a Hellfire missile shot from a drone.

Since 2002, the United States has been using drones to target terrorists. The technical name for a military drone is "unmanned combat aerial vehicle." People often use the shortened version—UCAV. (When such a plane carries no weapons, it is simply a UAV, or "unmanned aerial vehicle.")

The key word is "unmanned." Drones have no onboard pilots or other human crew. They fly strictly by remote control. The "pilots," or persons who control them, are often far away. In fact, they can be stationed anywhere on Earth.

This incredible technology sealed al-Awlaki's fate. The drone that killed him came from a secret American base in the Arabian peninsula. Its pilots sat in a room at another base inside the United States. They flew the drone into Yemen and fired the missile.

An AGM-114 Hellfire missile hangs on the rail of an US Air Force MQ-1L Predator Unmanned Aerial Vehicle.

Aircraft mechanics replace a multispectural targeting system ball on an MQ-1B Predator.

IN THE AIR: DRONES

A REVOLUTION IN WARFARE

Mere minutes following the mission, news of its success reached President Barack Obama. More than a year before, he had given a secret order. It was directed to U.S. intelligence agents. They are in charge of gathering information about enemies of the United States. Sometimes aided by the military, they also capture or destroy those enemies. Obama insisted that al-Awlaki be found and either captured or killed.

On hearing of the drone attack, the president addressed the American public. "The death of Awlaki is a major blow to al-Qaeda," he said. Obama went on to explain al-Awlaki's "lead role" in the terrorist group. Al-Awlaki had planned and directed "the efforts to murder innocent Americans," the president stated.

President Obama at a press conference on September 10, 2011.

President Obama and other U.S. leaders have come to rely on drones. Indeed, these lethal devices have recently changed the nature of warfare. "They represent a revolution in the idea of what combat is," *Time* magazine reporter Lev Grossman writes. "With drones," he adds, "the U.S. can exert force not only instantly . . ." it can also do it without "the risk of incurring American casualties." An expert on modern warfare, Peter Singer, agrees. He says that drones have altered "everything from tactics" to "overall strategy." They have changed "how leaders, the media, and the public" see and understand "this thing we call war."

CHAPTER TWO
A WEAPON Whose Time Had Come

THE MILITARY EXPERIENCE.

The first United States use of an unmanned plane, or drone, to attack and destroy a target occurred in 2002. Since that time, the use of war, or military, drones has skyrocketed. As a result, many people assume that such devices are a recent development.

The reality, however, is that UCAVs are far from new. They have been under development or in use for more than a century. The first known example appeared in August 1849. The Austrian military planned an assault on the Italian city of Venice. But the Austrians wanted to avoid heavy casualties on their side. So they outfitted several unmanned balloons with bombs.

The plan was to let the balloons drift over Venice. The bombs were rigged with fuses that would make them explode while over the city. A few of them did just that. But unexpected gusts of wind blew many others away. They drifted back over Austrian lines, causing the plan to backfire.

WORLD WAR I DRONE TESTING

The first winged airplanes designed as military drones were created during World War I (1914-1918). They were intended to shoot down German Zeppelins, enormous airships that floated upward because they were filled with lighter-than air gases. Fleets of piloted Zeppelins flew over and dropped bombs on London.

A house in London damaged during a raid by five German Zeppelins in October 1915

The German Zeppelin LZ 18 (L 2) at Berlin-Johannisthal in 1913

THE MILITARY EXPERIENCE.

To counter the Zeppelins, English inventor Archibald M. Low conceived of an unmanned airplane. He believed he could make it fly by sending it radio signals. One signal would order the drone to move up or down. Another would make it go left or right.

Low tested the initial prototype in 1917. It was a small craft made of wood and thin sheets of tin. An onboard thirty-five horsepower engine powered a propeller in the front.

A large group of British military officers gathered to watch the demonstration. But Low was embarrassed when the drone failed to perform as planned. It flew in circles and crashed not far from the launch point. Low was convinced that, given time, he could make the plane work. But only a few months after the test flight, the war ended. So Low's drone program shut down.

At that time few people knew about similar tests of military drones in the United States. During the war, an American inventor, Elmer A. Sperry, developed a small unmanned plane flown by remote control. Because it carried explosives, it came to be called an "air torpedo." Sperry and the U.S. Navy held secret tests of the device. They took place in a deserted area in central Long Island, New York. But like Low's program, the American version shut down when the war ended.

Elmer Sperry with another of his inventions, the remote controlled searchlight

IN THE AIR: DRONES

THE WORLD WAR II ERA

Nevertheless, inventors continued to experiment with *unarmed* drones. Major strides were made in the 1930s by Hollywood actor Reginald Denny. His hobby was building radio-controlled model planes.

In 1934, Denny founded a hobby shop that soon became the Radioplane Company. It developed drones for use as targets for training U.S. pilots. The planes worked so well that the U.S. Army took notice. In 1940 it awarded the company a major contract. The Radioplane Company turned out almost 15,000 target drones during World War II (1939-1945).

Meanwhile, during that conflict the U.S. Navy ran its own drone program. Dubbed Operation Anvil, its goal was to use unmanned craft to attack and destroy targets in Germany. The Navy gathered several B-24 bombers and loaded them with explosives. The plan was to use radio signals to guide the drones.

The B-24 Bomber

But remote control technology was still very primitive. So the program needed human pilots to get the planes in the air. They were supposed to parachute to safety soon after takeoff. This ambitious experiment failed, however. Operation Anvil shut down after several of its drones crashed.

FAST fact

Many of the bomb-filled drones tested for Operation Anvil in World War II crashed. The men who flew them during takeoff died. Among these brave test pilots was President John F. Kennedy's older brother, Joseph.

Lieutenant Joseph P. Kennedy, Jr.

The Kennedy family at the Kennedy Compound in Hyannis Port, Massachusetts (September 4, 1931). Seated left to right: Robert, John, Eunice, Jean (*on lap of*) Joe, Sr., Rose (*behind*) Patricia, Kathleen, and Joe, Jr. (*behind*) Rosemary. Ted had not been born yet.

IN THE AIR: DRONES

FACING A NEW REALITY

Yet Denny's target drones had been a resounding success. This kept interest in drone technology alive. In the decades following World War II, U.S. military engineers explored a different use for drones. The goal was to use them for surveillance, or spying on enemy countries.

The Central Intelligence Agency, or CIA, also saw the potential for spy drones. In the 1980s and 1990s, the CIA and Air Force made major strides in UAV technology. Advanced electronics, especially computers, vastly improved drone guidance systems.

Then 9/11 happened. And from that point, late in 2001, everything changed. It became clear that terrorists had become a major threat to the United States. U.S. military experts faced a new reality. Finding and killing small groups of fighters was not a job for large armies. Instead, targeting terrorists required a new approach to warfare. What was needed was a device that could both track down and kill a handful of individuals. With a bit of tinkering, that device already existed. It was the military drone, a weapon whose time had come.

World Trade Center
September 11, 2001

Captain Ryan Jodoi (*rear*) flies an MQ-9 Reaper while Airman First Class Patrick Snyder controls a full motion video camera at Kandahar Air Base, Afghanistan.

IN THE AIR: DRONES

Rise of Drones in the U.S. Military

A major development in the history of drones began in 2002. It was the rapid increase in their use by the U.S. military. Before that, most Air Force pilots flew traditional warplanes. They looked on drone pilots as unmanly geeks. No one envisioned that drones might be used in serious warfare. But that situation swiftly changed. Today more than sixty U.S. Air Force bases operate hundreds of military drones. The pilots send these lethal devices all over the world. In fact, in 2013 the Air Force had more drone pilots in training than pilots for conventional warplanes.

CHAPTER THREE

How DRONES and Their PILOTS Operate

THE MILITARY EXPERIENCE.

Looking at photos or films of a modern military drone can be misleading. These flying vehicles are petite and lightweight. They are also fairly smooth, featureless, and lacking in outer detail. But their smallness and seeming simplicity mask a far more complex reality. They carry millions of dollars worth of advanced equipment, including the world's most sophisticated imaging tools.

The fact that these planes are unmanned and remote-controlled can also give a false impression. Many people picture them working like the model planes available at hobby shops. That is, a single individual holds a small control box. He or she points a built-in antenna at the drone. Working a small lever, the person makes the plane take off, zoom around, and land.

In truth, the use of today's military drones is hugely more involved than that. To keep a single plane in service requires dozens of highly trained personnel. Together, they make it fly over a specific country, but in doing so they may be located all over the world. Drones also utilize a complex, very expensive system of orbiting communications satellites.

TWO MAIN DRONE TYPES

Thus, drones both carry their own technology and tap into outside technical systems. Some of the onboard technology is classified, or top secret. But a surprisingly large amount is public knowledge. It reveals that both of the two main types of U.S. military drones look physically similar. But each has its special features and uses.

One of these unmanned craft is the MQ-1B Predator. It has a wingspan of 55 feet (17 meters) and can cruise at up to 25,000 feet (7,620 meters). The MQ-1B is used mostly for extended missions involving long-term surveillance. Sometimes the military wants to keep an eye on an individual or group for weeks or months. So it sends an MQ-1B or similar-style drone to do the job.

IN THE AIR: DRONES

The other main U.S. military drone is the MQ-9 Reaper. It has a wingspan of 66 feet (20 meters) and can operate as high as 50,000 feet (15,240 meters). The Reaper is more often employed in short-term missions. These include verifying the nature of a target and then moving right in to destroy it. Of the two vehicles, the Reaper is faster. It can fly at 230 miles per hour (370 kilometers per hour).

The Reaper is also the more lethal of the two main drone types. It carries a deadly payload of four AGM-114 Hellfire missiles. (In comparison, the MQ-1B carries two Hellfire missiles.) These weapons are laser guided. That makes them extremely accurate. Once fired, a Hellfire can travel up to five miles before demolishing its target.

MQ-9 Reaper

THE MILITARY EXPERIENCE.

FAST fact

The edges of a predator drone's wings have numerous tiny holes. Through these, a special liquid seeps out from small storage tanks inside. That fluid breaks down ice that can form on the wings in cold air.

PAINTING THE TARGET

Despite these differences, the two drone types have many features and abilities in common. Each is equipped with an impressive array of advanced electronic equipment. This includes sophisticated radar and satellite communications gear. There is also state-of-the-art GPS (Global Positioning System). It is similar to the version in most modern cars, only far more complex. A drone's GPS allows it to hone in on any spot on Earth's surface.

The most remarkable devices these drones carry are their cameras and other sensors. Most are located in the nose assembly. They make it possible for a drone to find and target its prey. The cameras are extraordinarily powerful. They often "bring war straight into a pilot's face," noted journalist Elizabeth Bumiller says.

The drone's camera is very powerful.

IN THE AIR: DRONES

One drone pilot agrees. "I used to drop bombs from a flying airplane," he explains. "I could not see the faces of the people." With drones, by contrast, it is like being on the ground. According to another pilot, it feels like sneaking up on the target. "It freaks you out," he says. "You feel less like a pilot than a sniper."

While the pilot studies these visual images, the drone's sensors are also at work. They "calculate wind speed, direction, and other battlefield variables," one researcher explains. "This process is known as 'painting the target.' Once a target is painted," the drone "can unleash its own missiles to destroy the target." Or it can send the collected data "to other aircraft or ground forces so they can destroy it."

Dealing with the Bad Days

Every time a drone pilots fires on a target, he or she knows it is part of the job. "I feel no emotional attachment to the enemy," one pilot says. "I have a duty, and I execute the duty." Still, he points out, sending a missile to blow up a human being is not fun. It is very different than destroying a bad guy in a video game. "There was good reason for killing the people that I did," another pilot stated. "But you never forget about it. It never just fades away." Such feelings are magnified in cases where a pilot accidentally kills innocent bystanders. Some pilots become so upset about it that they need counseling. To that end, journalist Elizabeth Bumiller reports, the Air Force keeps "chaplains and medics just outside drone operation centers to help pilots deal with the bad days."

Pilots perform function checks after launching an MQ-1 Predator.

AN "ENGINEERING SPECTACLE"

These drone pilots do their job inside a building called a ground control station (GCS). Such stations are usually far from the battlefield. One pilot, Colonel D. Scott Brenton, controls a Reaper drone. It is located 7,000 miles (11,270 kilometers) away in Afghanistan. He watches the images sent by the drone on large screens. When he sends an order to the plane, the signal speeds to a satellite. From there, it rushes—at the speed of light—to the drone.

Thus, the pilots sit safely, far from their targets. Yet most still feel a surge of tension when they prepare to destroy a target. Brenton claims the hair on the back of his neck stands up. Each pilot reacts to that part of the job differently.

But all of them agree that recent advances in drone technology have been incredible. "They are truly an engineering spectacle," one expert puts it. They contain "the best of mechanical, electronics, and software technology. There just might be a day when today's generation tells their grandchildren" that warplanes were once "manned by human pilots."

A pilot remotely flies the MQ-9 Reaper at Kandahar Airfield, Afghanistan.

WARNING

خطرنا الخطر

RESTRICTED AREA

It is unlawful to enter this area without permission of the Installation Commander.
Sec. 21, Internal Security Act of 1950; 50 U.S.C. 797

In this installation all personnel and the property under their control are subject to search.

Use of deadly force authorized

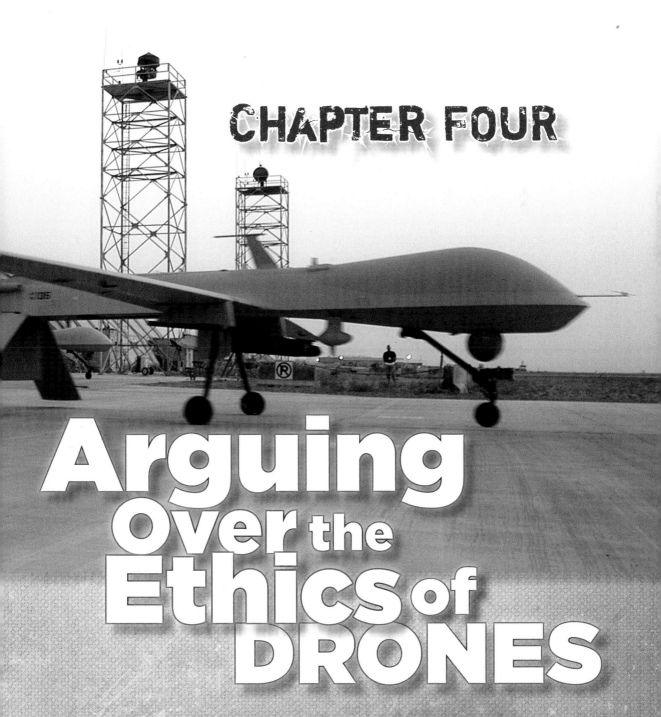

CHAPTER FOUR

Arguing Over the Ethics of Drones

THE MILITARY EXPERIENCE.

U.S. military drones began to make an impact on warfare in 2002. George W. Bush was the first president to authorize them. His successor, Barack Obama used them, too. In fact, in his first four years in office Obama okayed six times as many drone strikes as Bush did in eight years.

At first, most Americans did not have strong opinions about the use of drones. They saw them as just another weapon to use against terrorists. But over time, certain ethical and legal questions arose.

Today, those questions spark a good deal of public discussion and debate. Noted journalist David E. Sanger explains why such debate is timely. Military drones might well "change the way we fight the wars of the future," he says. So legal and moral questions about them "need to be part of the public conversation."

TOO MUCH COLLATERAL DAMAGE?

A major issue raised in these debates is the number of civilians killed by drones. Accidental civilian deaths in warfare are often called "collateral damage." Critics of military drones claim these weapons create too much of that damage.

To back up this claim, they cite various recent studies, conducted by reputable organizations. Among them were Stanford University's and New York University's law schools. Another is London's Bureau of Investigative Journalism (TBIJ). These groups researched civilian deaths in the Middle East and Asia caused by U.S. drones.

The results of the TBIJ study were announced in 2012. It said that from 2004 to 2012 U.S. drone strikes killed at least 2,562 people in Pakistan. Of these, the researchers said, at least 474 were civilians. Moreover, 176—more than a third—were children.

IN THE AIR: DRONES

The researchers came to a conclusion: U.S. drone strikes are too destructive to civilians. They called for "a significant rethinking of current U.S. targeted killing" by drones. "U.S. policy-makers," they went on, "cannot continue to ignore evidence of the civilian harm."

In response, President Obama defended his military use of drones. He argued that "drones have not caused a large number of civilian casualties." Obama and CIA leaders said that the figures gathered in the studies were misleading. Many of the supposed civilians killed were secretly aiding the terrorists.

Some U.S. leaders offer another argument supporting drone use. Namely, some of these civilian deaths cannot be helped. Collateral damage is regrettable, they say. But it occurs in all wars.

American citizens hold a banner during a peace march in Tank, Pakistan.

THE MILITARY EXPERIENCE.

FAST fact

In March 2013, Senator Rand Paul of Kentucky filibustered congress for twelve hours, questioning whether or not the government could use drones against American citizens on U.S. soil. Paul's filibuster drew both criticism and support from Republicans and Democrats, illustrating how divided this issue is.

Senator Rand Paul

IN THE AIR: DRONES

DRONE USE UNJUST?

People have raised other ethical questions about military drones. For example, some think it is unfair to use them in warfare. Unlike traditional warplanes, they say, drones have no onboard pilots. So in a drone war, American combatants have no risk of dying. In contrast, the other side loses hundreds or thousands of fighters. This approach to war, these critics suggest, is unjust.

Well-known Historian David Bell disagrees. He defends using military drones. "We don't know," he says, if anybody condemned "the inventor of the bow and arrow." Was he "a dishonorable coward who refused to risk death in a hand to hand fight?" Also, Bell points out, some medieval people denounced the crossbow. They felt it "allowed poorly skilled archers to kill honorable knights from safe cover." This was not honorable, they said. It would be fairer if all soldiers fought as equals.

Yet such arguments are nonsense, Bell argues. Armies and nations always look for ways to gain the advantage over enemies. "Using technology to strike safely at an opponent," he adds, "is as old as war itself."

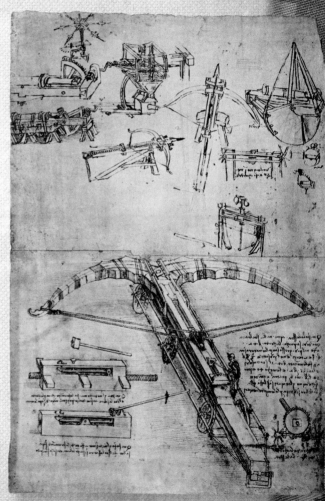

A Leonardo da Vinci sketch of a giant crossbow

A soldier communicates with the pilot of an MQ-1 Predator while it is being prepared to launch for a test flight in Iraq.

THE MILITARY EXPERIENCE.

DRONES HERE TO STAY

Still another question has been raised about military drone use. It is whether U.S. drones should kill American citizens. A number of Americans and Europeans criticized the drone strike on Anwar al-Awlaki in 2011. They agreed he was a bad man. But he was also an American, they said. So he should have been captured and tried in an American court. According to this view, killing al-Awlaki was illegal.

But President Obama and his advisors see it differently. They insist that killing people like al-Awlaki is both justified and legal. To back their argument, they point to a 2012 secret memo. It came from the U.S. Justice Department. The memo says an American can be targeted if he or she poses an immediate violent threat to the United States. Thomas E. Ricks, an expert on military affairs, has examined all these arguments. The drone program "may have its faults," he admits. But it does get rid of a lot of people who want to destroy America. So, "for better or for worse," he says, "drones are here to stay."

An MQ-1 Predator armed with AGM-114 Hellfire missile

IN THE AIR: DRONES

Should Drones Target Thieves?

Using military drones has begun to spread beyond warfare. In June 2011, a North Dakota sheriff's office searched for three men. Accused of theft, they were believed to be heavily armed. The search widened to hundreds of square miles. So the sheriff decided on a novel strategy. He called a local Air Force base and asked to borrow a military drone. The plane easily found the suspects. Moreover, its cameras showed they were actually unarmed. This marked the first time a drone led to the arrest of U.S. citizens. A number of individuals and organizations called this domestic use of military drones disturbing.

CHAPTER FIVE
The FUTURE Use of MILITARY DRONES

In the recent past, military drones have been largely successful against terrorists. Award-winning journalist Thomas E. Ricks says that drone strikes have "crippled al-Qaeda's leadership." Moreover, this has occurred in the space of only a few years. Also, he says, fear of drones will probably deter many young men from joining such groups.

Hicks and other military experts say these recent successes are crucial. They are likely, they believe, to affect the future of drones in warfare. This is partly because Americans deeply dislike and fear terrorism. And they will likely accept almost any weapon that keeps them safe from it.

There are other reasons that military drone use will continue to increase. One, says military expert David Cortright, is that they "are inexpensive." So they "seem to make the waging of war less costly." Also, "they allow leaders to conduct military operations without risking the lives of U.S. soldiers." In turn, that draws less "public disapproval."

For these reasons, U.S. military leaders have come to embrace drone use. The proof is the massive recent expansion of their drone programs. In 2000, a year before 9/11, the U.S. military had around ninety drones. At the time, it planned to build only two hundred more by 2010. Now, CNN reporter Josh Levs writes "the United States has 8,000 drones."

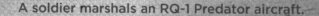
A soldier marshals an RQ-1 Predator aircraft.

THE MILITARY EXPERIENCE.

Contractors load an AGM-114 Hellfire missile onto an MQ-1 Predator in Iraq.

IN THE AIR: DRONES

A REVOLUTION OF MILITARY AFFAIRS?

More importantly, Levs says, the U.S. military "has a robust plan for using more and more" drones "in the future." For example, the Air Force bought two dozen new Reaper drones in early 2013. It also planned to buy more than four hundred more in the following three or four years. That did not count the drones ordered by the Army, Navy, and CIA that year.

Moreover, it is not just the number of drones that is rising. The number of missions they fly is also increasing. The U.S. Air Force estimates that its drones took part in five missions each day in 2012. It expects that number to rise to close to seventy a day by 2016. That means "more countries with drones flying over them," military observer Nick Turse writes.

Turse and other experts also predict that drone technology will not stand still. Rather, it will continue to progress at a rapid pace. By 2020, they say, sending a single drone for each mission will be outdated. Just one small control crew will be able to fly several drones at once.

Forecasting what drones will do even further in the future is difficult. But Air Force planners have tried. "Unmanned aircraft will be able to refuel each other by 2030," they say. "As technology advances, machines will automatically perform some repairs in flight."

These predictions may actually be conservative, some experts say. Future drones may well have advanced onboard computer brains. This may give them the incredible ability to fly missions without human pilots. This is like the *Star Wars* robot R2D2 flying a spacecraft on its own. One Air Force pilot is certain these things will happen. He calls drone technology "a revolution of military affairs." It will happen, he believes, through "the conscious application of automated technology."

THE MILITARY EXPERIENCE.

Hidden in Plain Sight

One major aspect of advancing drone technology is making these machines smaller. The smaller they are, the thinking goes, the better. They will then be much harder to detect and destroy. In fact, military researchers are trying to shrink drones to the size of birds and insects.

"We're looking at how you hide in plain sight," says engineer Greg Parker. Scholar Peter W. Singer has written about the possible uses of these tiny robots. He smilingly calls them "spy flies" and "bugs with bugs." They will have tiny cameras and sensors, he predicts. An early version of such miniature drones has already been built. The company that designed it calls it a "hummingbird drone." It can fly at eleven miles per hour (eighteen kilometers per hour) and land on a windowsill.

The Nano Hummingbird surveillance aircraft

IN THE AIR: DRONES

A DIFFERENT WORLD

Many U.S. leaders are confident about these advances in drones. They believe they will help defend the country for decades to come. But some military experts point out a possible problem. CNN investigator Peter Bergen sums it up concisely. Only a few years ago, he says, America "had a virtual monopoly on drones. Not anymore."

Bergen is referring to a growing presence of military drones in foreign arsenals. Indeed, in 2013, more than seventy nations had at least a few drones. Among them were the United Kingdom, Australia, China, Russia, and Israel. Only a few of their planes were then armed with weapons. But many of these countries are rapidly developing that capability.

Israel's IAI Heron

American leaders are not bothered that allies such as the UK and Israel have drones. What disturbs them is that some adversaries claim to have them. One is Iran, situated in the heart of the Middle East. Its leaders have been openly hostile to the United States. Some military experts worry about Iran using drones to attack America or its allies. Another future concern is that various nations will become involved in "drone wars."

Future relations among nations may therefore become more complicated. Having a larger army and navy may no longer ensure a country's safety. With drones, a smaller, less powerful nation could do serious damage to stronger enemies. Many prophecies about drones in the future may or may not come true. But one thing is certain. They have already changed the nature of warfare forever.

FAST fact

In 2010, China announced its creation of twenty-five kinds of military drones. Some of them carry missiles, a Chinese official claimed.

A New Zealand soldier prepares a UAV 'KAHU' to launch.

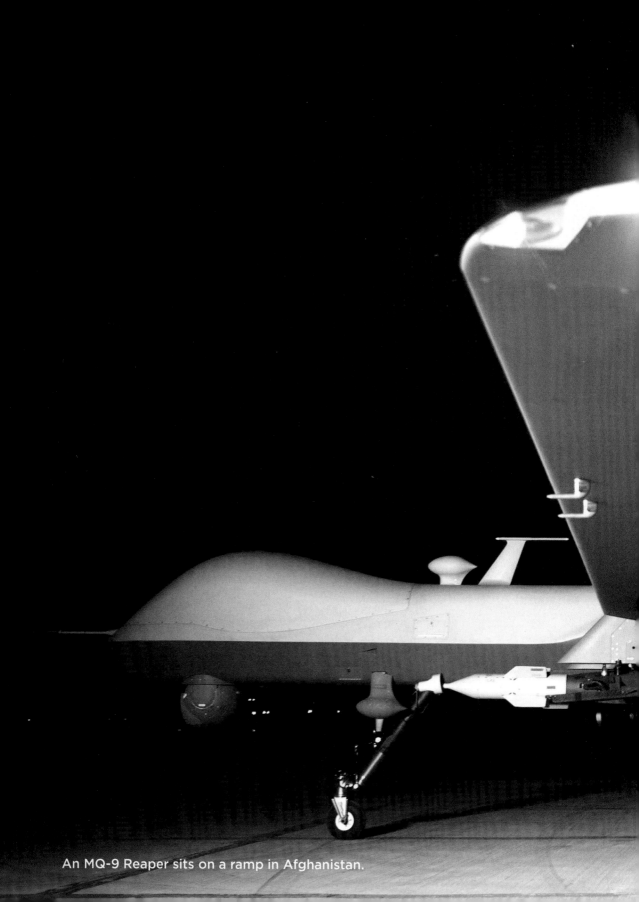
An MQ-9 Reaper sits on a ramp in Afghanistan.

Sources

Chapter 1: Fireball in the Yemeni Mountains

p. 17, "The death of Awlaki is . . ." Mark Mazzetti et al, "Two-Year Manhunt Led to Killing of Awlaki in Yemen," http://www.nytimes.com/2011/10/01/world/middleeast/anwar-al-awlaki-is-killed-in-yemen.html?_r=1&.

p. 17, "They represent a revolution . . ." Lev Grossman, "Drone Home," http://www.time.com/time/magazine/article/0,9171,2135132,00.html.

p. 17, "everything from tactics . . ." Ibid.

Chapter 3: How Drones and Their Pilots Operate

p. 32, "bring war straight . . ." Elizabeth Bumiller, "A Day Job Waiting for a Kill Shot a World Away," http://www.nytimes.com/2012/07/30/us/drone-pilots-waiting-for-a-kill-shot-7000-miles-away.html?pagewanted=all&_r=0.

p. 33, "I used to drop bombs . . ." David E. Sanger, *Confront and Conceal: America's Secret Wars and Surprising Use of American Power* (New York: Crown, 2012), 257.

p. 33, "It freaks you out . . ." Ibid.

p. 33, "calculate wind speed . . ." Robert Valdes, "How the Predator UAV Works: In Battle," http://science.howstuffworks.com/predator4.htm.

p. 33, "I feel no emotional attachment . . ." Bumiller, "A Day Job Waiting for a Kill Shot a World Away."

p. 33, "There was a good reason . . ." Ibid.

p. 33, "chaplains and medics . . ." Ibid.

p. 34, "They are truly . . ." V. Shalen Prevas, "Aerial Assassins: Drones," http://readanddigest.com/what-is-a-drone/.

Chapter 4: Arguing Over the Ethics of Drones

p. 38, "change the way we . . ." Sanger, *Confront and Conceal*, 270.

p. 39, "a significant rethinking . . ." CNN Wire Staff, "Drone Strikes Kill, Maim and Traumatize Too Many Civilians, U.S. Study Says," http://www.cnn.com/2012/09/25/world/asia/pakistan-us-drone-strikes.

p. 39, "drones have not caused . . ." Sanger, *Confront and Conceal*, 251.

p. 41,   "We don't know . . ." David Bell, "In Defense of Drones: A Historical Argument," http://www.newrepublic.com/article/politics/100113/obama-military-foreign-policy-technology-drones#.

p. 41,   "Using technology . . ." Ibid.

p. 44,   "may have its faults . . ." Thomas E. Ricks, "Drones: Here to Stay," http://ricks.foreignpolicy.com/posts/2012/11/07/drones_here_to_stay.

Chapter 5: The Future Use of Military Drones

p. 48,   "crippled al-Qaeda's leadership," Thomas E. Ricks, "Are the Strategic Costs of Obama's drone Policy Greater than the Short-term Gains?," http://ricks.foreignpolicy.com/posts/2012/06/27/are_the_strategic_costs_of_obama_s_drone_policy_greater_than_the_short_term_gains_0.

p. 48,   "are inexpensive," David Cortright, "The Scary Prospect of Global Drone Warfare." http://www.cnn.com/2011/10/19/opinion/cortright-drones.

p. 48,   "The United States has . . ." Josh Levs, "CNN Explains: U.S. Drones," www.cnn.com/2013/02/07/politics/drones-cnn-explains.

p. 51,   "has a robust plan. . ." Ibid.

p. 51,   "more countries with drones . . ." Nick Turse, "Drone Wars: Pentagon's Future with Robots, Troops, Part 3," http://www.cbsnews.com/8301-215_162-57445801/drone-wars-pentagons-future-with-robots-troops/?pageNum=3.

p. 51,   "Unmanned aircraft will . . ." Joe Pappalardo, "The Future for UAVs in the U.S. Air Force," http://www.popularmechanics.com/technology/aviation/military/4347306.

p. 51,   "a revolution of . . ." Ibid.

p. 52,   "we're looking at . . ." Elizabeth Bumiller, "War Evolves With Drones, Some Tiny as Bugs," www.nytimes.com/2011/06/20/world/20drones.html.

p. 53,   "had a virtual monopoly . . ." Levs, "CNN Explains: U.S. Drones."

Glossary

civilian: A person who is not in the armed forces.

classified: Top secret.

collateral damage: The military and legal term for *civilian* harmed or killed in combat.

combatant: A soldier or other fighter.

drone: An unmanned flying vehicle.

GPS (Global Positioning System): A network of orbiting satellites that allow people to quickly compute their exact position on Earth's surface.

intelligence: Information about one's adversaries.

memo: A note, letter, or message.

monopoly: Domination or sole control of something.

predator: An animal, person, or drone that hunts and kills its prey.

prototype: The first example or test subject of a human-made object or idea.

strategy: Plans for how to accomplish a goal.

surveillance: Closely watching or spying on a person, group, or nation.

tactics: Methods of or approaches to accomplishing a job or mission.

technology: The application of scientific principles to tools, work, industry, and/or commercial activities.

UCAV: Unmanned Combat Aerial Vehicle; the technical name for a drone.

Zeppelin: A large airship held aloft by an enormous bag filled with hydrogen or some other lighter-than-air substance.

Bibliography

Books

David, Jack. *Predator Drones*. Minneapolis: Bellwether Media, 2007.

Nagelhout, Ryan. *Drones*. New York: Gareth Stevens, 2013.

Rudy, Lisa Jo. *Micro Spies: Spy Planes the Size of Birds!* New York: Children's Press, 2007.

Sanger, David E. *Confront and Conceal: America's Secret Wars and Surprising Use of American Power.* New York: Crown, 2012.

Springer, Paul J. *Military Robots and Drones*. Santa Barbara, CA: ABC-CLIO, 2013.

Turse, Nick. *Terminator Planet: The First History of Drone Warfare, 2001-2050*. Create Space, 2012.

Web sites

David Bell, "In Defense of Drones: A Historical Argument"
http://www.newrepublic.com/article/politics/100113/obama-military-foreign-policy-technology-drones#

Elizabeth Bumiller, "War Evolves With Drones, Some Tiny as Bugs"
www.nytimes.com/2011/06/20/world/20drones.html

David Cortright, "The Scary Prospect of Global Drone Warfare"
http://www.cnn.com/2011/10/19/opinion/cortright-drones

Cora Currier, "Everything We Know So Far About Drone Strikes"
http://www.propublica.org/article/everything-we-know-so-far-about-drone-strikes

Lev Grossman, "Drone Home"
http://www.time.com/time/magazine/article/0,9171,2135132,00.html

Josh Levs, "CNN Explains: U.S. Drones"
http://www.cnn.com/2013/02/07/politics/drones-cnn-explains

Robert Valdes, "How the Predator UAV Works: In Battle"
http://science.howstuffworks.com/predator4.htm

YouTube, "Drone Controllers at Work"
http://www.youtube.com/watch?v=7Dxn_qEs_p8

Index

AGM-114 Hellfire missile, 14, *14–15*, 31, *44–45, 50–51*
air torpedo, 22
Al-Awlaki, Anwar, 11, *11*, 12–14, 17, 44
al-Qaeda, 11, 13, 17, 48

balloons, unmanned, 20
B-24 Bomber, 23, *23*
Bush, George W., 38

China and drones, 54
CIA and drones, 25, 39
collateral damage, 38–39

Denny, Reginald, 23, 25
drones, *18–19, 36–37*
    and changes in warfare, 17, 25, 38, 41, 54
    equipment in, 30, 32–33
    ethical issues, 38, *39*, 40–41, 44
    future use of, 48, 51, 53–54
    history of, 20, 22–23
    missions, 14, 17, 31, 51
    remote pilot operation of, 14, 27, 28–29, 30, 33–34, *34*, 35
    and technology, 14, 25, 26, *28–29*, 30, 32–33, *32*, 34, 51–52
    types of, 30, 32, 51–52, *52*
    uses of, 14, 23, 25, 30, 45

Fort Hood shooting, 12, *12*

GCS (Ground Control Station), 34
GPS (Global Position System) units, 32

Hasan, Nidal Malik, 12, *12*
hummingbird drone, 52

IAI Heron (Israeli drone), *53*
Iraq and drones, 54

KAHU UAV (New Zealand drone), *54–55*
Kennedy, Joseph P., Jr., 24, *24*
Khan, Samir, 13

London, bombing of, 20, *20–21*
Low, Archibald M., 22

missiles, 14, 31, 33, 54
MQ-1 Predator, *34, 42–43, 44–45*
MQ-1B Predator, *4–5, 8–9, 16,* 30–31
MQ-1L Predator, *14–15*
MQ-9 Reaper, *2, 26,* 31, *31, 35, 56–57*

Nano Hummingbird surveillance aircraft, *52*

Obama, Barack, 17, *17*, 38, 44
Operation Anvil, 23, 24

Paul, Rand, 40, *40*
predator drones, 32

Radioplane Company, 23
Reaper drones, 51
remote controlled spotlight, *22*
remote control techonology, 23, 30
RQ-1 Predator, *46–47, 48–49, 50–51*

September 11 terrorist attacks, 11, 25, *25*
Sperry, Elmer, 22, *22*

TBIJ study on drones, 38–39
terrorism, 11, 14, 17, 25, 48

UAV (Unmanned aerial vehicle), 14
UCAV (Unmanned combat aerial vehicle), 14, 20
Unmanned Aerial Vehicle, *14–15*
U.S. Air Force and drones, 25, 27, 30, 44–45, 51
U.S. Army and drones, 23, 48, 51
U.S. Navy and drones, 22–23, 51

warfare, drone affected changes, 17, 25, 38, 41, 54
World War I, 22
World War II, 23, 24

Zeppelin LZ18, *20–21*
zeppelins, 20, 22

Photo Credits

All images used in this book that are not in the public domain are credited in the listing that follows.

2: Courtesy of U.S. Air Force
8-9: Courtesy of U.S. Air Force
10 (globe): Courtesy of TUBS
10-11: Courtesy of yeowatzup
11 (al-Awlaki): Courtesy of Muhammad ud-Deen
16: Courtesy of U.S. Air Force
18-19: Courtesy of U.S. Air Force
25: Courtesy of Kevinalbania
26: Courtesy of U.S. Air Force
28-29: Courtesy of U.S. Air Force
31: Courtesy of U.S. Air Force
34: Courtesy of U.S. Air Force
35: Courtesy of U.S. Air Force
36: Courtesy of U.S. Air Force
39: Associated Press
40: Associated Press
41: Courtesy of Library of Congress
42-43: Courtesy of U.S. Air Force
46-47: Courtesy of U.S. Air Force
48: Courtesy of U.S. Air Force
50: Courtesy of U.S. Air Force
54-55: Courtesy of New Zealand Defence Force
56-57: Courtesy of U.S. Air Force